50 THINGS TO SEE WITH A TELESCOPE

A YOUNG STARGAZER'S GUIDE

John A. Read

Formac Publishing Company Limited
Halifax

For my boys, Isaac and Oliver.

Formac Publishing Company Limited recognizes the support of the Province of Nova Scotia through the Department of Communities, Culture and Heritage. We are pleased to work in partnership with the Province of Nova Scotia to develop and promote our cultural resources for all Nova Scotians. We acknowledge the support of the Canada Council for the Arts, which last year invested $153 million to bring the arts to Canadians throughout the country. This project has been made possible in part by the Government of Canada.

Cover design: Tyler Cleroux
Cover image: Istock

Library and Archives Canada Cataloguing in Publication

Read, John A., author
    50 things to see with a telescope : a young stargazer's guide / John A. Read.
-- New edition.

ISBN 978-1-4595-0536-0 (hardcover)

    1. Astronomy--Observers' manuals--Juvenile literature. 2. Astronomy--Amateurs' manuals--Juvenile literature. 3. Telescopes--Amateurs' manuals--Juvenile literature. I. Title. II. Title: Fifty things to see with a telescope.

QB63.R395 2018   j520   C2018-903014-3

Published by:
Formac Publishing
Company Limited
5502 Atlantic Street
Halifax, NS, Canada
B3H 1G4
www.formac.ca

Distributed in Canada by:
Formac Lorimer Books
5502 Atlantic Street
Halifax, NS, Canada
B3H 1G4

Distributed in the US by:
Lerner Publisher Services
1251 Washington Ave. N.
Minneapolis, MN, USA
55401
www.lernerbooks.com

Printed and bound in Canada.
Manufactured by Friesens Corporation in Altona, Manitoba, Canada in July 2018.
Job # 245950

ACKNOWLEDGEMENTS
Special thanks to my editors Kurtis Anstey, Kara Turner, Jennifer Read and David M.F. Chapman. Thank you to everyone at Formac for your excellent collaboration on this project.

PHOTO CREDITS
Telescope view source files for deep-sky objects were constructed from actual photos taken by the author, either using his personal four-inch refractor, twelve-inch Dobsonian and eight-inch Dobsonian, or using the following remote observatories: Abbey Ridge Observatory (owned by Dave Lane), and the Burke-Gaffney Observatory at Saint Mary's University, Halifax. Exceptions include M1, imaged by Kurtis Anstey; Comet C/2013 US10, imaged by Dave Lane; and M81, M82, and the Double Cluster, by Stuart Forman.

Other images used include: images from NASA which follow NASA's photo usage guidelines; image of Comet 67P/Churyumov Gerasimenko from ESA/Rosetta/NAVCAM, CC BY-SA IGO 3.0; image of Celestron FirstScope Dobsonian compliments of Celestron; image of Explore Scientific FirstLight refractor compliments of Explore Scientific; image of Andromeda on back cover by Adam Evans courtesy of Wikimedia; background images in interior from Shutterstock.

Star maps used in this book were sourced using Stellarium, an open-source stargazing program. These maps were then customized for the purpose of this book. Several of artist Johan Meuris constellation images from Stellarium are included in this book and usage rights can be found here: artlibre.org/licence/lal/en/.

BLE OF CONTENTS

Using This Book

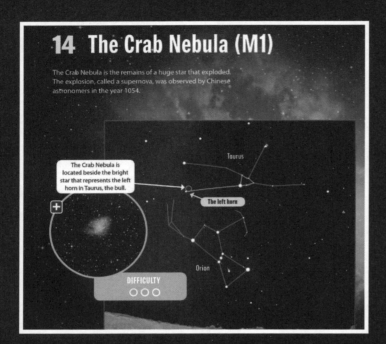

14 The Crab Nebula (M1)

The Crab Nebula is the remains of a huge star that exploded. The explosion, called a supernova, was observed by Chinese astronomers in the year 1054.

Taurus

The Crab Nebula is located beside the bright star that represents the left horn in Taurus, the bull.

The left horn

Orion

DIFFICULTY

This book is designed as an introduction to stargazing. Most sections introduce a star pattern (a constellation or asterism) that can be identified without a telescope, with arrows directing you to the telescope targets within that part of the sky.

The little blue circle on the map represents an estimation of how much sky you might see through your telescope.

These round windows on nearly every page show how the object will look through your telescope in perfectly dark skies. Note: galaxies and nebulae (giant clouds of gas and dust) will require extremely dark skies to appear as they do in these images.

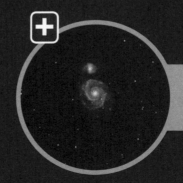

DIFFICULTY

If you see this symbol, the object can be viewed with binoculars.

Planets (items 42–49) appear to wander through the ecliptic (the path the Sun travels across the sky) and require software to locate on any given night. The stargazing software "Stellarium" is free and can be downloaded at *www.Stellarium.org* or from the app store.

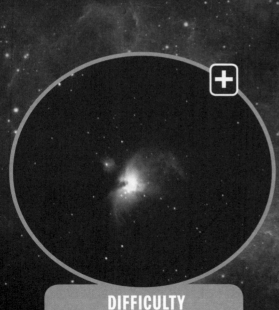

DIFFICULTY
○ ○ ○

Attached to each telescope view window is a measure of how challenging an object is to observe. Level 1 can be found with ease (assuming the object is above the horizon). Level 2 requires some patience, while level 3 requires extremely dark skies or, in the case of Uranus and Neptune, the use of stargazing software.

The Sky Above Us

Humans have gazed up at the sky since before the dawn of civilization. The movements of the Sun and planets along with the fixed positions of the stars helped people know when to plant their crops and how to navigate the seas.

Nearly every culture on Earth grouped stars into patterns and gave them names. The Greeks named one group Orion the Hunter. Meanwhile, Chinese astronomers included the same stars as two of 28 Mansions. Hindi astronomers called these stars the Deer. The most popular star pattern, which we today recognize as the Big Dipper, has had dozens, if not hundreds of names throughout history. The Inuit people of what is now northern Canada at one point called it *Tukturjuit*, the Caribou. In eastern Europe, it was called the Great Wagon, while Arabian societies viewed these stars as a coffin.

The Caribou

Greater Bear

Stellar Facts

A star pattern within a constellation is called an asterism. Asterisms have common names like the Big Dipper, the Diamond or the Teapot.

Island universes

We live within a collection of stars called the Milky Way Galaxy. Up until the early 1900s, many astronomers thought the Milky Way (which contains about 300 billion stars) was the entire universe. Although other galaxies could be seen through almost any telescope, scientists did not know how distant they were, assuming they were clouds of gas within our galaxy. But in 1920, an astronomer named Heber Curtis argued otherwise, calling these objects "island universes." In 1923, Edwin Hubble proved Curtis right, calculating the distance to the Andromeda Galaxy and many others, proving once and for all that these nebulae were not nebulae at all, but individual galaxies each containing billions of stars.

Objects that are not comets

A French comet hunter named Charles Messier created a list of 110 fuzzy objects he saw through his telescope. Messier didn't know what they were at the time, but he could tell that they weren't comets. We now know the following as deep-sky objects: open star clusters, globular star clusters, nebulae and galaxies. The objects are named according to Messier's initial "M" and their number in this catalogue. Messier's list is now the primary target list for amateur astronomers. Most deep-sky objects mentioned in this book are included in Messier's list.

M31 – The Andromeda Galaxy

M1 – The first object in Messier's list of objects that aren't comets

M103 – Open Cluster

M20 – Trifid Nebula

Dark Skies

The objects in this book can be located in the night sky in the northern hemisphere as long as it is the correct season and the sky is clear. However, some galaxies, nebulae and globular clusters require dark or very dark skies. How dark are your skies? Use the images below as a guide.

POOR SKY	FAIR SKY	DARK SKY	VERY DARK SKY
In a town, or during a full moon.	Suburban skies, 10 kilometres from the nearest town..	Country skies, 20 kilometres from the nearest town.	50 kilometres from the nearest town

The Whirlpool Galaxy (M51) viewed under different sky conditions

How many stars can you see?

Though a telescope, you're able to see millions of stars. Without a telescope, there are fewer than 10,000 stars visible, and only about 2,500 are visible at any one time. Near a town, or when the moon is full, you'll only be able to see a few hundred stars.

In a city, you might only see a dozen! How many stars are there in the observable universe? We can estimate by multiplying the average number of stars in a galaxy by the number of visible galaxies. The total is about one septillion stars (1,000,000,000, 000,000,000,000,000), although astronomers believe the actual number is much higher than this.

Each blob in this image from the Hubble Space Telescope is a galaxy containing hundreds of billions of stars.

Hopefully you know that Earth revolves around the Sun. This fact has a fascinating consequence in astronomy. As Earth orbits the Sun, the nighttime side faces a different part of the sky — the stars overhead at night in the winter are overhead during the day summer. For this reason, the constellations and targets in this book are ordered by seaso

Not all stars rise and set. Many stars in the northern sky can be seen all year from the northern hemisphere. (If you are south of the equator — in Australia, for example — it's the southern stars that do not rise and set.) When you look up at the night sky for any length of time, you'll notice that the stars appear to rotate around the North Star. A complete rotation occurs about once every day, as Earth spins, and about once every year, as Earth revolves around the Sun.

Objects in the night sky that never rise or set but appear to circle the North Star are referred to as "circumpolar." We'll explore many of these objects in Chapter 1.

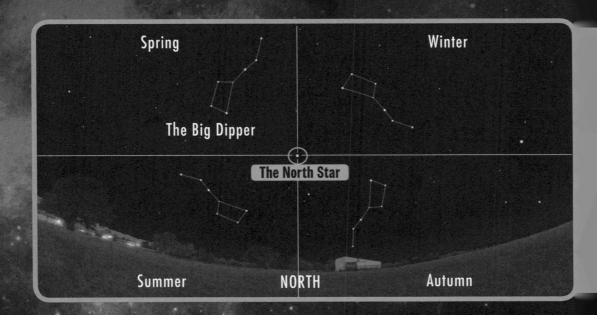

Choosing a Telescope

Amateur astronomy is a challenging hobby, even for an adult. The targets in this book are beyond the range of toy telescopes. In general, the best telescopes for beginners are either Dobsonians or refractors on solid alt/az (up/down–left/right) mounts. Here are a few things you'll want to look for in a telescope:

Avoid telescopes on flimsy or camera tripods. These telescopes may be marketed to kids, but they are extremely challenging to point at objects in space.

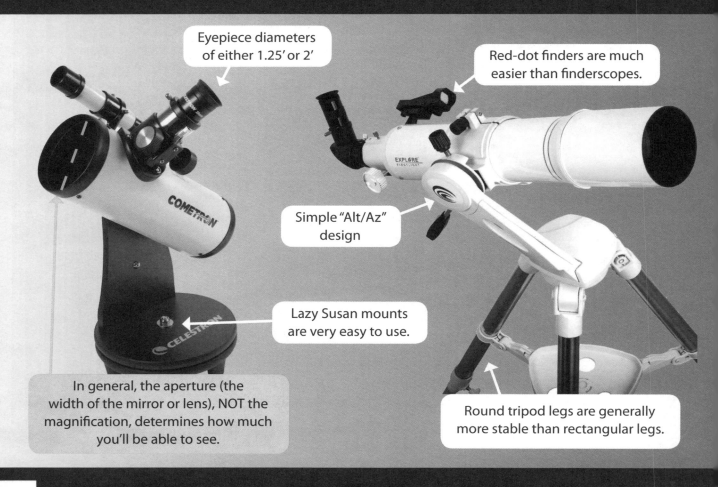

Eyepiece diameters of either 1.25' or 2'

Red-dot finders are much easier than finderscopes.

Simple "Alt/Az" design

Lazy Susan mounts are very easy to use.

In general, the aperture (the width of the mirror or lens), NOT the magnification, determines how much you'll be able to see.

Round tripod legs are generally more stable than rectangular legs.

Parts of a Telescope

Refractors, like the telescope on the left, use lenses to magnify distant objects.

Reflectors, like the Newtonian telescope on the right, use mirrors to direct light into the eyepiece.

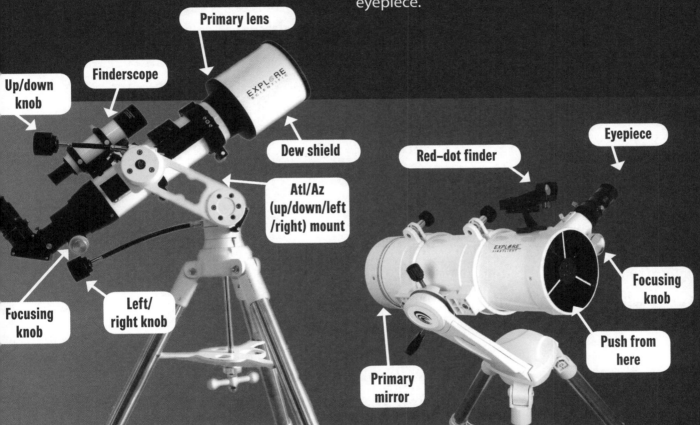

Primary lens

Finderscope

Up/down knob

Dew shield

Atl/Az (up/down/left /right) mount

Focusing knob

Left/ right knob

Eyepiece

Red–dot finder

Focusing knob

Push from here

Primary mirror

Equatorially mounted (EQ) telescopes (not shown) are designed to track the Earth's rotation along a single tilted axis. They have additional features that can be challenging for young kids.

Getting Started

Setting up your telescope

When you set up your telescope, be sure to follow the manual closely or find a video online specific to your telescope, and follow the instructions. Try to set up your telescope on solid, level ground and not a deck. Vibrations from walking on the deck will travel through the telescope and make the image bounce. It's important to have a clear view of most of the sky unobstructed by trees or buildings and away from artificial sources of light.

Once you have the telescope assembled, make sure it is working properly by testing to see if it can point in all directions. Ensure the telescope and mount stays in place when you let go.

Choosing an eyepiece

Most beginner telescopes come with two eyepieces, one with a larger lens (more glass) than the other. The eyepiece with the larger lens is the one you want to use most of the time. Only use the smaller one when you want to zoom in on a target like a planet. You'll find you won't need to zoom in very often because the most important thing is light gathering, not magnification.

Many telescopes come with a lens called a "3x Barlow" or "2x Barlow." These devices are designed to be placed between the eyepiece and the telescope to triple or double the magnification. However, this also makes your telescope much more difficult to aim and focus, and most of the time the Barlow attachment is unnecessary.

Barlow lens (use sparingly)

Large eyepiece (left) and small eyepiece (right) (Ideally use the larger one)

A filter may also have been included with your telescope. Filters thread into the bottom of the eyepiece before the eyepiece is set into the telescope. The filter, which may be labelled "Moon" or "Polarized" is designed to reduce brightness and see more details when observing the Moon.

Moon (or polarizing) filter

Focusing your telescope

In order to see anything through your telescope, it must be in focus. To do this, point the telescope at the Moon or a bright star. Then, twist the focusing knob until the image of the Moon is

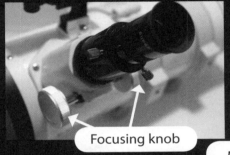

Focusing knob

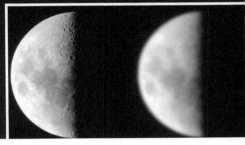

Moon in focus (left) and out of focus (right)

sharp or a bright star is as small as you can make it.

Aligning your telescope

For a telescope to work properly, the finderscope (or red-dot finder) must be aligned so that it points at exactly the same place as the telescope. To do this, point the telescope at a bright star. Twist the alignment knobs on the finder scope until the star is centred in both the finderscope and the telescope. If using a red-dot finder, the device must also be turned on.

Red-dot finder

Adapting your eyes to the dark

Viewing objects like galaxies, nebulae and globular clusters also requires you to prepare your eyes. It takes about 30 minutes to adapt your eyes to see these objects. This means you can't look at car headlights, porch lights or cell phones. It also means no flashlights (unless covered with red cellophane) and no looking at the Moon.

Star Hopping

To find any object in the night sky, you'll have to plot a route! Imagine you're giving directions to the nearest store. You might say, "Turn right at the traffic signal and left at the stop sign." The same strategy works in the night sky. A seasoned stargazer might say: "Follow the pointer stars to the North Star. Then hop over to Cassiopeia — you'll find the Dragonfly Cluster near the bottom left star in the W." This may sound confusing, but it will come naturally as you learn the constellations and bright stars.

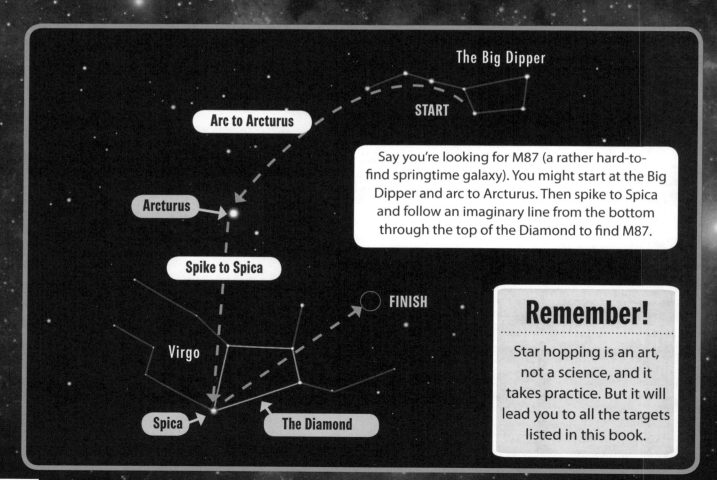

The Big Dipper

Arc to Arcturus

START

Arcturus

Spike to Spica

FINISH

Virgo

Spica

The Diamond

Say you're looking for M87 (a rather hard-to-find springtime galaxy). You might start at the Big Dipper and arc to Arcturus. Then spike to Spica and follow an imaginary line from the bottom through the top of the Diamond to find M87.

Remember!

Star hopping is an art, not a science, and it takes practice. But it will lead you to all the targets listed in this book.

CHAPTER 1
Year-Round Objects

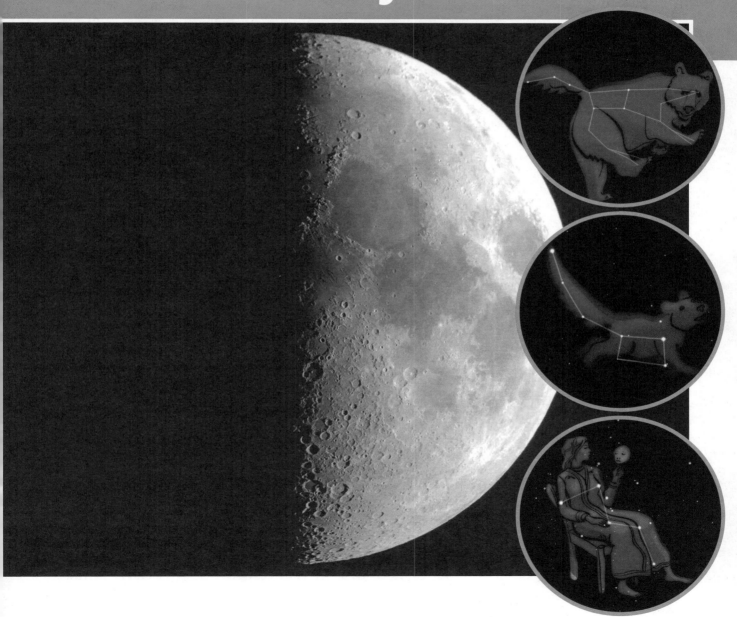

01 The Moon

The Moon completes a full cycle through its phases about once every 29 days. Each night, the Moon's phase is slightly different.

The Full Moon as viewed through a small telescope or binoculars.

DIFFICULTY
○ ○ ○

The Moon viewed at the same time each evening.

Night 7

First-Quarter Moon

Gibbous Moon

Crescent Moon*

Night 14

New Moon

Night 1

Full Moon

Eastern Horizon

Southern Horizon

Western Horizon

After the Full Moon, the Moon "wanes" through the following phases: Waning Gibbous, Third Quarter, Waning Crescent and then back to New Moon.

*As viewed from the Northern Hemisphere

Earth

Moon

This is a scale image of the distance between the Earth and the Moon.

– –

The average distance between the Earth and the Moon is 384,000 kilometres.

Lunar eclipse

Earth's shadow

Solar eclipse

Moon's shadow

Stellar Facts

The reason we don't have eclipses every month is because the Moon's orbit is slightly tilted. This means that most months, the shadows miss!

02 The Big Dipper & The Surfboard Galaxy

The Big Dipper is the most recognizable shape in the night sky. It is circumpolar, meaning that it stays above the horizon for most people living in the northern hemisphere. The stars in the Big Dipper make great targets to explore with your telescope. In very dark skies, try to find the Surfboard Galaxy (M108) close to the bowl of the Dipper.

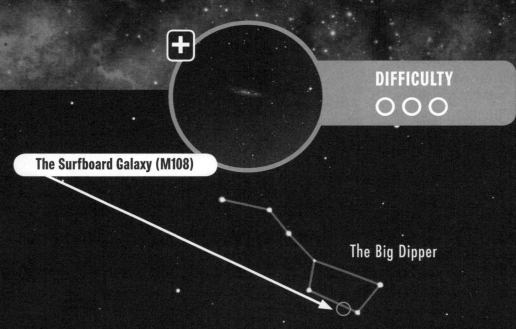

DIFFICULTY

○ ○ ○

The Surfboard Galaxy (M108)

The Big Dipper

03 Mizar & Alcor

Mizar and Alcor (nicknamed the Horse and Rider) make up the centre of the handle of the Big Dipper. Both Mizar and Alcor are visible without a telescope. What makes them interesting is that through a telescope, you'll notice that Mizar is actually two stars!

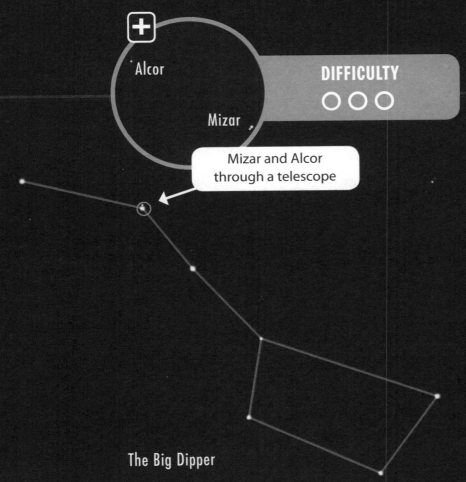

Alcor

Mizar

DIFFICULTY

Mizar and Alcor through a telescope

The Big Dipper

04 The Whirlpool & Pinwheel Galaxies

These two galaxies, located near the Big Dipper, make great targets for winter, spring and summer (they are a bit low in the sky in the fall). If you're near a town or city, if the Moon is up or if you have not adapted your eyes to the dark, the Pinwheel may be invisible, but in dark skies it's a beautiful sight.

Galaxies often look dim and blurred, mainly due to imperfect sky conditions. Astronomers call these views "beautiful smudges."

DIFFICULTY
○ ○ ○

The North Star

The Pinwheel Galaxy (M101) is visible through a small telescope in extremely dark skies.

The Big Dipper

The Whirlpool Galaxy (M51) is brighter than M101 and can be found in mildly light-polluted skies.

DIFFICULTY
○ ○ ○

05 The North Star (Polaris)

The entire northern sky appears to move around the North Star — it stays in the same place all year. It is called Polaris because it stays so close to the celestial pole. Many people think it is the brightest star in the sky, but in fact it's number 48. (Sirius, found in Canis Major, wins the prize for the brightest star.)

Stellar Facts

This star was important for sailors navigating at sea. The angle between this star and the horizon, multiplied by 69, provides the sailor their distance (in miles) from the equator!

The North Star

Find the North Star by following these two "pointer" stars in the Big Dipper.

The Big Dipper

Through a telescope, you may be able to see a companion star, Polaris B.

Polaris A

Polaris B

DIFFICULTY
○ ○ ○

06 The Little Dipper & Bode's Nebula

It can be a challenge to identify the Little Dipper because its stars are quite dim. Start by finding the North Star at the end of the handle and work your way over to the cup.

Bode's Nebula, a spiral galaxy, and the Cigar Galaxy (M82) near the Big Dipper are visible through a small telescope almost every clear night. You should be able to see both galaxies at once.

Stellar Facts

The Little Dipper is a nickname for the constellation Ursa Minor, which means "lesser bear."

The Little Dipper

The Big Dipper

The North Star

Cigar Galaxy (M82)

Bode's Nebula (M81)

DIFFICULTY

○ ○ ○

07 The Big W & Cluster M103

The Big W (or Cassiopeia) is always found on the opposite side of the North Star from the Big Dipper. Knowing how to find the Big W will lead to several other targets in this book, such as the Andromeda Galaxy and the Dragonfly Cluster. Open star cluster M103 is found within the Big W.

DIFFICULTY
○ ○ ○

Open star cluster M103 was discovered in 1781.

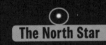
The North Star

The Big W
(Cassiopeia)

The Big Dipper

08 The Dragonfly (NGC 457)

Looking more closely at the Big W, you will find plenty of interesting star patterns. The most fun cluster to see through a telescope is the Dragonfly. Recently, this cluster has become known as the E.T. Cluster, named after the alien from the Steven Spielberg movie, *E.T.: The Extra-Terrestrial*.

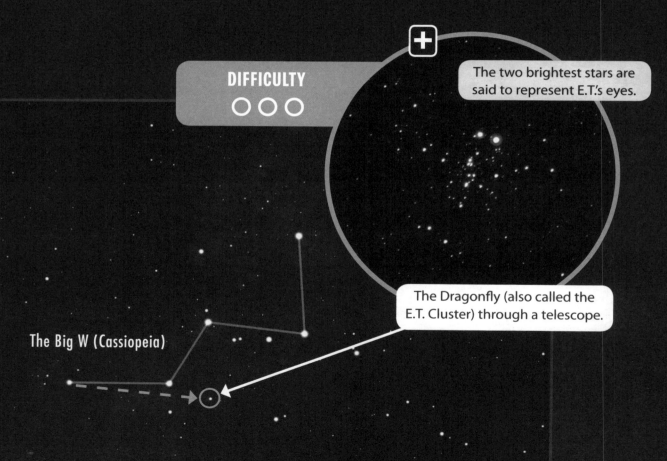

DIFFICULTY

○ ○ ○

The two brightest stars are said to represent E.T.'s eyes.

The Big W (Cassiopeia)

The Dragonfly (also called the E.T. Cluster) through a telescope.

09 Kemble's Cascade

In the dim constellation Camelopardalis lies a beautiful chain of stars named after Father Lucien Kemble, a Canadian priest. Because Camelopardalis is difficult to identify, you'll need to use the Big W (Cassiopeia) as a guide.

Kemble's Cascade through a telescope or binoculars.

DIFFICULTY
○ ○ ○

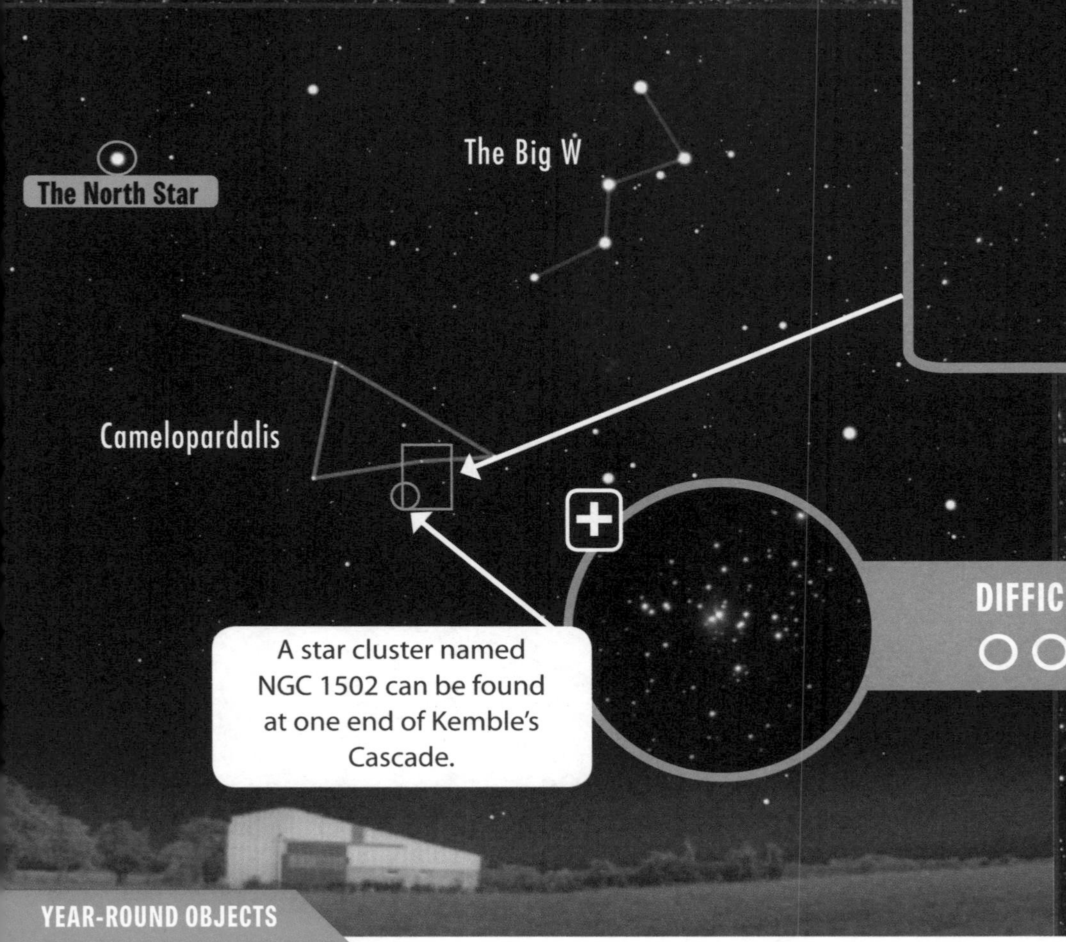

The Big W

The North Star

Camelopardalis

DIFFICULTY
○ ○ ○

A star cluster named NGC 1502 can be found at one end of Kemble's Cascade.

CHAPTER 2
Winter Objects

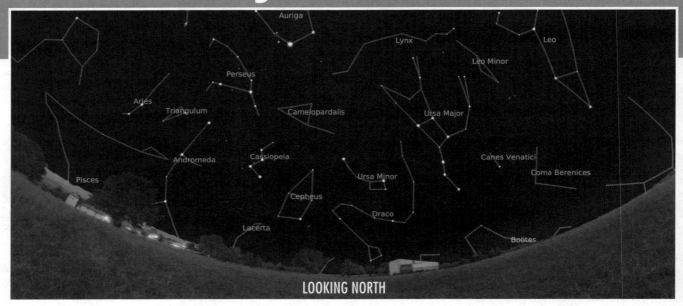

LOOKING NORTH

LOOKING SOUTH

10 Orion (The Hunter)

Orion is the most prominent winter constellation. It is easily recognized by the three stars that make up Orion's Belt. The red star near the top of the constellation is named Betelgeuse, while the bottom-right star, a blueish-white star, is named Rigel.

DIFFICULTY
○ ○ ○

Betelgeuse

Rigel

Orion's Belt

Stellar Facts

Orion is a hunter from Greek mythology.

STELLAR FACT: This constellation is home to the famous Horsehead Nebula. This object is outside the range of the small telescope, but it makes a great target for astrophotographers.

The Great Nebula in Orion is arguably the most brilliant nebula in the sky and also the easiest to find. Located just below Orion's Belt, it can be spotted without a telescope as a smudge of light in Orion's Sword.

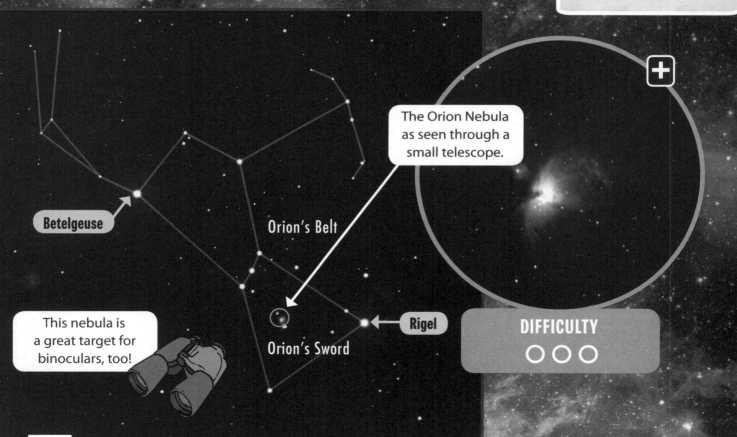

The Orion Nebula as seen through a small telescope.

Betelgeuse

Orion's Belt

Rigel

Orion's Sword

This nebula is a great target for binoculars, too!

DIFFICULTY
○ ○ ○

12 The Twins (Gemini) & Cluster M35

Gemini (or the Twins, as you can see by their constellation lines) is found near Orion in the winter. You can find this constellation by locating the two top stars, Pollux and Castor. Near the foot of the right twin is open star cluster M35.

Stellar Facts

Gemini is the location of one of the biggest meteor showers of the year.
The Geminids occur annually in mid-December.

Pollux

Castor

Gemini
(The Twins)

Orion

Open star cluster M35 is 2,800 light-years from Earth.

DIFFICULTY
○ ○ ○

13 Canis Major & Cluster M41

In Greek mythology, Canis Major (which means "Greater Dog") is following Orion, the hunter from the previous section. This constellation contains Sirius, also known as the "Dog Star." Canis Major is also home to open cluster M41, located in the middle of the constellation.

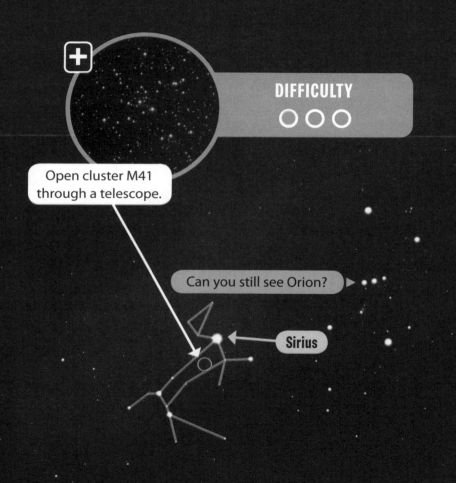

DIFFICULTY

Open cluster M41 through a telescope.

Can you still see Orion?

Sirius

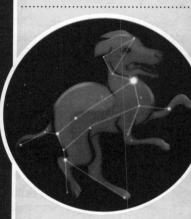

14 The Crab Nebula (M1)

The Crab Nebula is the remains of a huge star that exploded. The explosion, called a supernova, was observed by Chinese astronomers in the year 1054.

The Crab Nebula is located beside the bright star that represents the left horn in Taurus, the bull.

Taurus

The left horn

Orion

DIFFICULTY

WINTER OBJECTS

15 The Hyades (in Taurus)

Taurus means "bull" in Latin. It's a prominent winter constellation. The brightest star in Taurus is Aldebaran, which lies at one corner of the Hyades, the star cluster at the centre of the constellation.

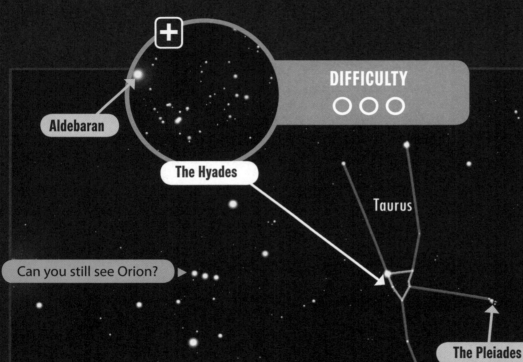

DIFFICULTY
○ ○ ○

Aldebaran

The Hyades

Taurus

Can you still see Orion?

The Pleiades

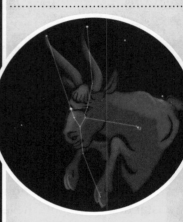

16 The Pleiades (M45)

This open cluster of stars is easily visible without a telescope in late autumn evenings and throughout the winter. Due to its shape, many people mistakenly believe this is the Little Dipper. Without a telescope, only six or seven stars are visible; but with a telescope, you'll see hundreds!

Through a telescope or binoculars, the Pleiades appear as a brilliant cluster of dozens of stars.

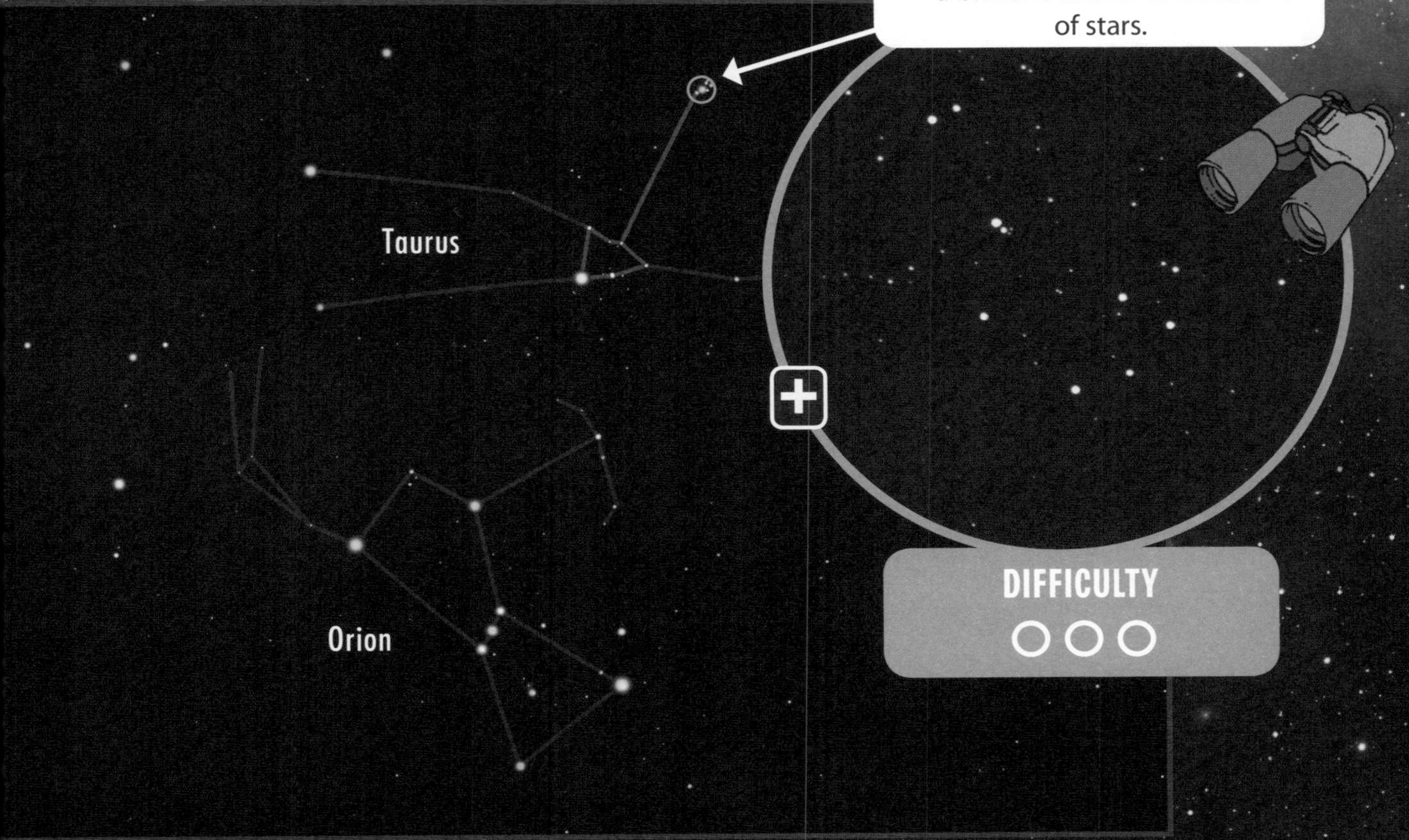

Taurus

Orion

DIFFICULTY

17 The Starfish Cluster & The Cloaking Warbird

Some star clusters may not look like much but, just like constellations, people have imagined patterns within them and given them names. M38 is known as the Starfish Cluster. I like to call M37 the Cloaking Warbird, after the Romulan starship from *Star Trek*!

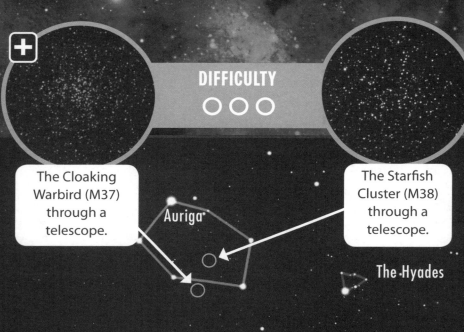

DIFFICULTY
○ ○ ○

The Cloaking Warbird (M37) through a telescope.

The Starfish Cluster (M38) through a telescope.

Auriga

The Hyades

Orion

18 Perseus & Spiral Cluster M34

Perseus lies between the Big W and Taurus. It is most famous for the Perseid meteor shower that occurs in mid-August. However, in the summer, Perseus doesn't rise until after midnight, so you'll have to stay up late to see the best shooting stars. Cluster M34 can be found close to Perseus.

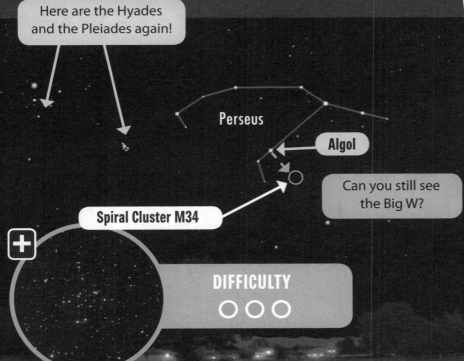

Here are the Hyades and the Pleiades again!

Perseus

Algol

Spiral Cluster M34

Can you still see the Big W?

DIFFICULTY
○ ○ ○

19 The Winter Hexagon & The Satellite Cluster

During the winter, a star pattern called the Winter Hexagon will orient you, helping you to identify the nearby constellations. The hexagon is formed by joining the stars Rigel, Aldebaran, Capella, Pollux, Procyon and Sirius.

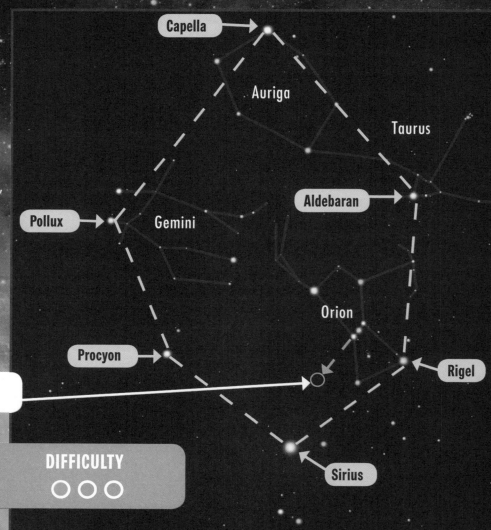

Capella

Auriga

Taurus

Pollux

Gemini

Aldebaran

Orion

Procyon

Rigel

Sirius

Satellite Cluster NGC 2244 through a telescope.

DIFFICULTY
○ ○ ○

CHAPTER 3
Spring Objects

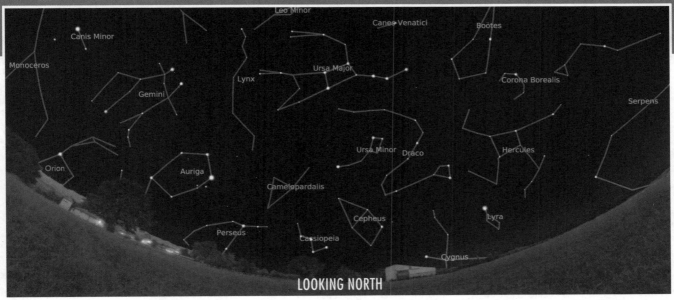

LOOKING NORTH

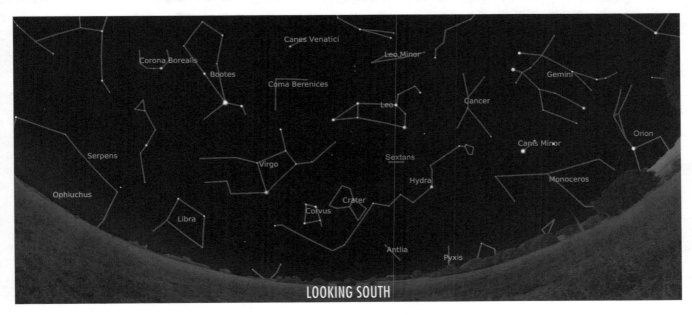

LOOKING SOUTH

20 Corona Borealis & Cluster M5

This small but beautiful constellation rests across from the handle of the Big Dipper. Corona Borealis means "northern crown" in Latin. The constellation boasts seven bright stars, arcing through a curve like knights at a round table.

The "ice cream cone" (Boötes)

The Diamond in Virgo

Corona Borealis

Arcturus

This bright star is officially called Alphekka, but it also goes by the name Gemma, which means "jewel" in Latin.

Unukalhai

Globular cluster M5 — a grouping of thousands of stars. Though visible in the eastern sky during the spring, M5 will be visible throughout the summer, too.

DIFFICULTY
○ ○ ○

38

SPRING OBJECTS

21 Boötes & Cluster M3

The constellation Boötes looks like a giant ice cream cone located next to the Big Dipper, though traditionally this constellation is depicted as a herdsman. The brightest star is Arcturus, the third-brightest star in the night sky.

Stellar Facts

In Greek mythology, Boötes represents Arcas, son of Zeus.

The Big Dipper

Boötes

Arcturus

Globular cluster M3

DIFFICULTY
○ ○ ○

22 The Diamond & The Sombrero Galaxy

You may want to lie on your back to find Virgo. Start by identifying its brightest star, Spica. To do this, start with the Big Dipper then arc to Arcturus, a red star, then spike (or speed) to Spica.

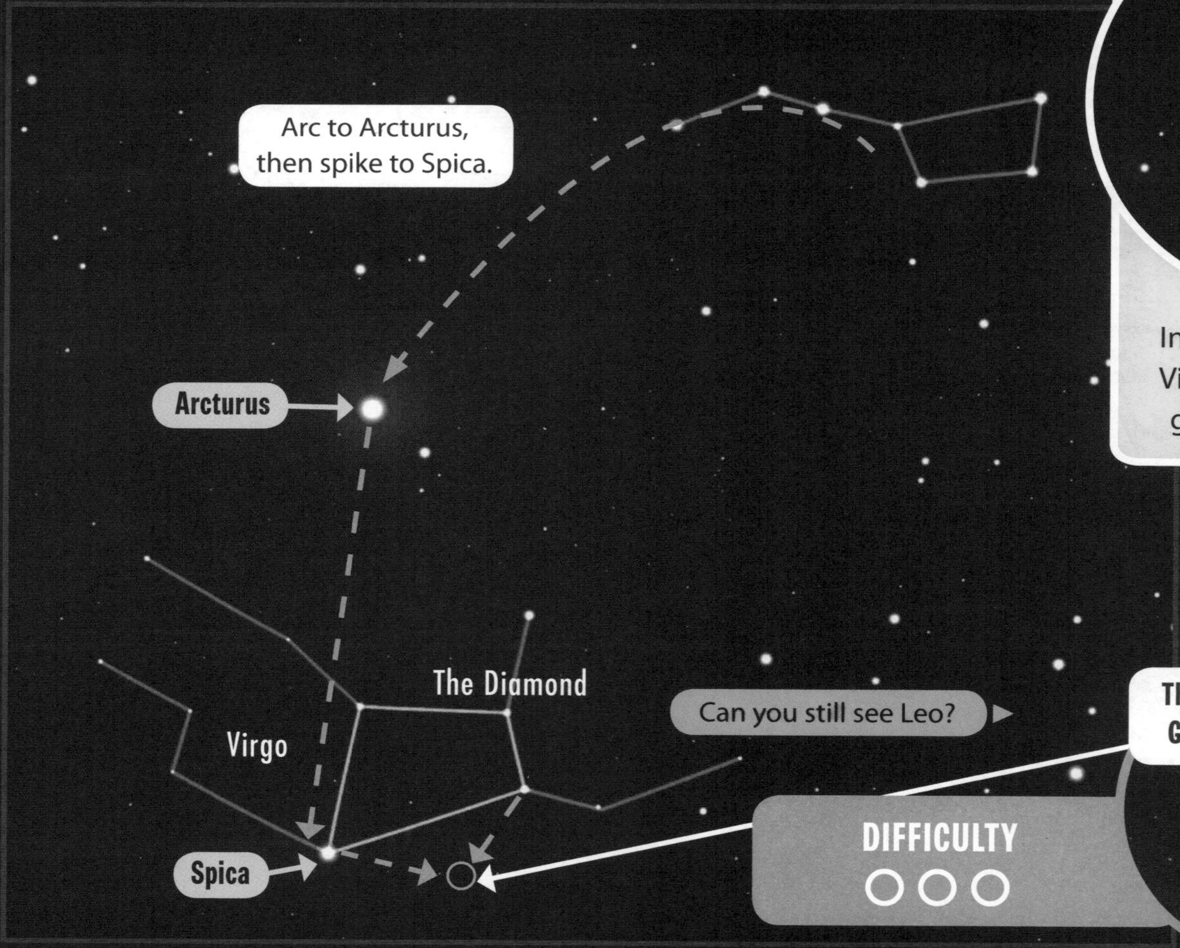

Arc to Arcturus, then spike to Spica.

Arcturus

The Diamond

Can you still see Leo?

Virgo

The Sombrero Galaxy M104

Spica

DIFFICULTY
○ ○ ○

23 Leo & The Hamburger Galaxy (NGC 3628)

Leo, the Lion, is a spring constellation, that to me looks more like a mouse. Some people recognize Leo by a question mark pattern known as the Sickle. The brightest star in this constellation is Regulus. Nicknamed the Hamburger Galaxy, NGC 3628 is part of the "Leo Triplet" — a group of galaxies viewable together through a small telescope.

Leo, the Lion

Leo

The Sickle

Regulus

Here are Pollux and Castor in Gemini, hiding low in the springtime sky.

The Hamburger Galaxy (NGC 3628) is visible only in extremely dark skies.

DIFFICULTY
○ ○ ○

24 Cancer & The Beehive

The stars in the constellation Cancer aren't very bright, making it a difficult constellation to identify. However, in the centre of Cancer lies an object called the Beehive (M44), a popular target for small telescopes and binoculars.

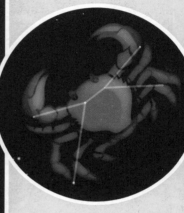
Regulus →

Cancer

Pollux and Castor

The Beehive (M44), an open star cluster, almost looks like a constellation within a constellation. It is a great target for binoculars.

Procyon

DIFFICULTY

○ ○ ○

CHAPTER 4
Summer Objects

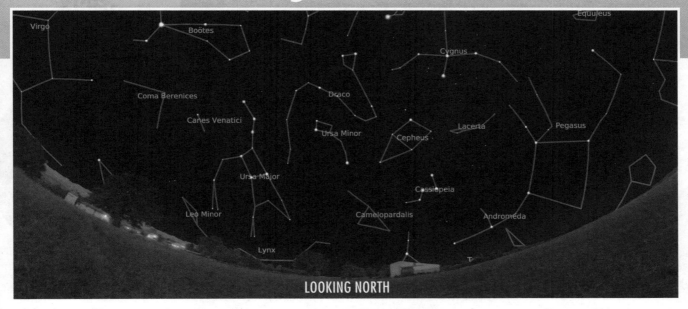

LOOKING NORTH

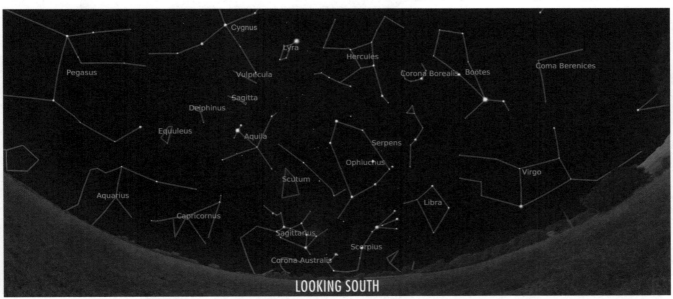

LOOKING SOUTH

25 Scorpius & Cluster M4

During the summer, Scorpius rises just above the southern horizon. It is easily identified by the claw and the bright red star, Antares. The top portion of the constellation forms a star pattern (asterism) sometimes called the Rake, while the bottom portion is sometimes referred to as the Longshoreman's Hook.

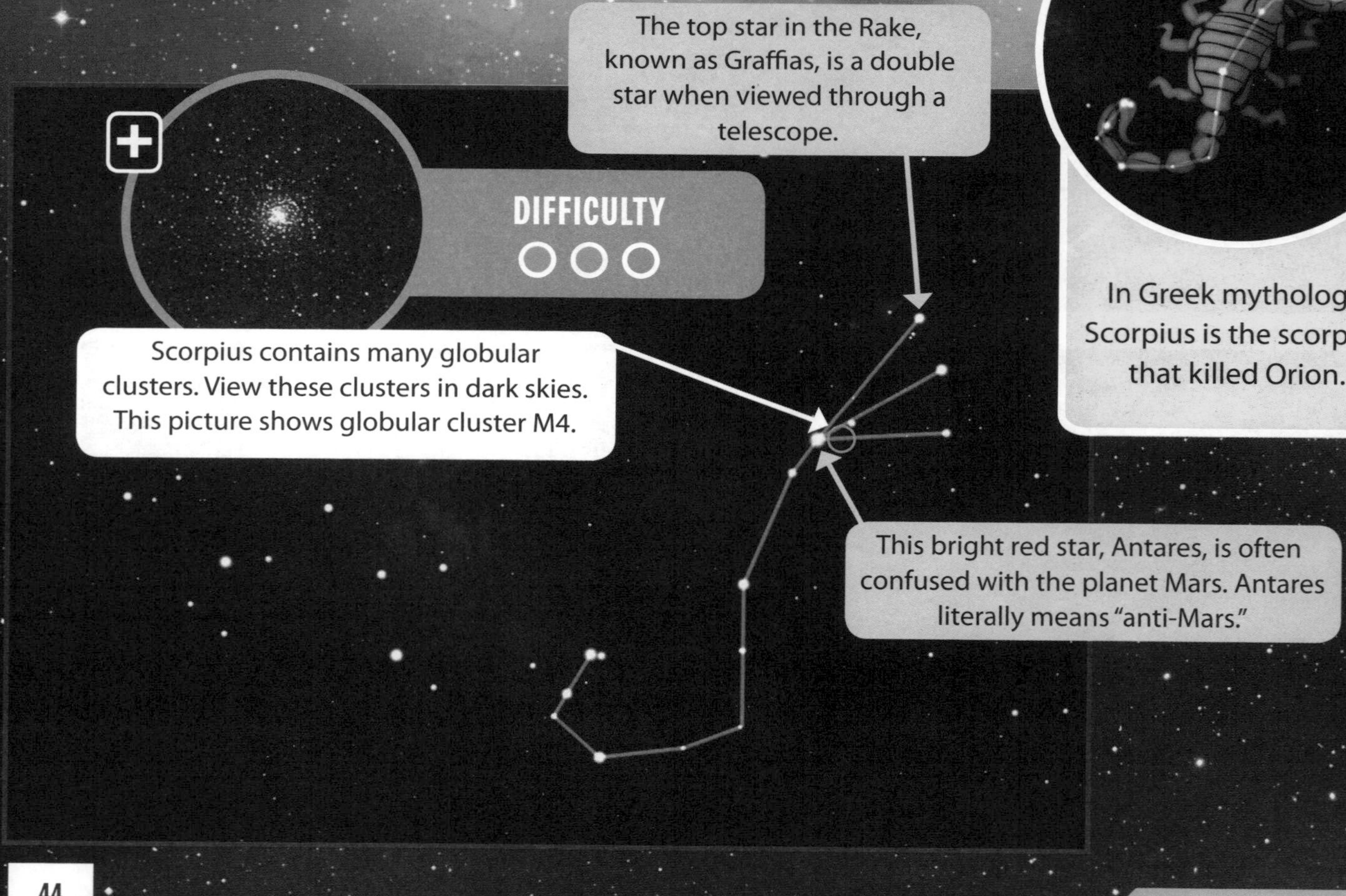

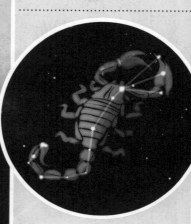

The top star in the Rake, known as Graffias, is a double star when viewed through a telescope.
DIFFICULTY
Scorpius contains many globular clusters. View these clusters in dark skies. This picture shows globular cluster M4.
This bright red star, Antares, is often confused with the planet Mars. Antares literally means "anti-Mars."

Stellar Facts

In Greek mythology, Scorpius is the scorpion that killed Orion.

26 The Omega Nebula & Ptolemy's Cluster

Just to the left of Scorpius rests an asterism called the Teapot, a star pattern within the constellation Sagittarius. This part of the sky is teeming with deep-sky objects, and is a wonderful area to explore with binoculars. Here are two objects you can find close to the Teapot.

The Omega (or Swan) Nebula (M17)

DIFFICULTY
○ ○ ○

The Teapot

Ptolemy's Cluster is a great target for binoculars, too!

Ptolemy's Cluster (M7)

DIFFICULTY
○ ○ ○

27 Cluster M22 & The Butterfly (M6)

Globular cluster M22 (found near the lid of the Teapot) and the Butterfly Cluster (found between the Teapot and Scorpius's tail) make great targets for binoculars! When viewing the Butterfly through a telescope, you'll need to use your imagination as you picture wings and antennas forming from arcing patterns of stars.

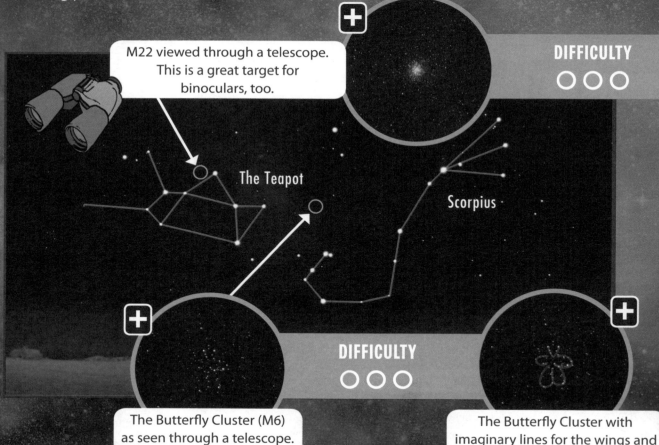

M22 viewed through a telescope. This is a great target for binoculars, too.

DIFFICULTY
○ ○ ○

The Teapot

Scorpius

DIFFICULTY
○ ○ ○

The Butterfly Cluster (M6) as seen through a telescope.

The Butterfly Cluster with imaginary lines for the wings and antennas.

28 Clusters M10 & M12 in Ophiuchus

Sitting above Scorpius and the Teapot, below Hercules and to the right of the Summer Triangle, Ophiuchus is known for its lack of central bright stars. It does, however, contain plenty of deep-sky treasures, including M9, M10, M12, M14 and M107 (all globular clusters).

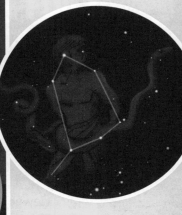
DIFFICULTY
○ ○ ○

Globular cluster M12

Ophiuchus

Globular cluster M10

DIFFICULTY
○ ○ ○

Scorpius

The Teapot

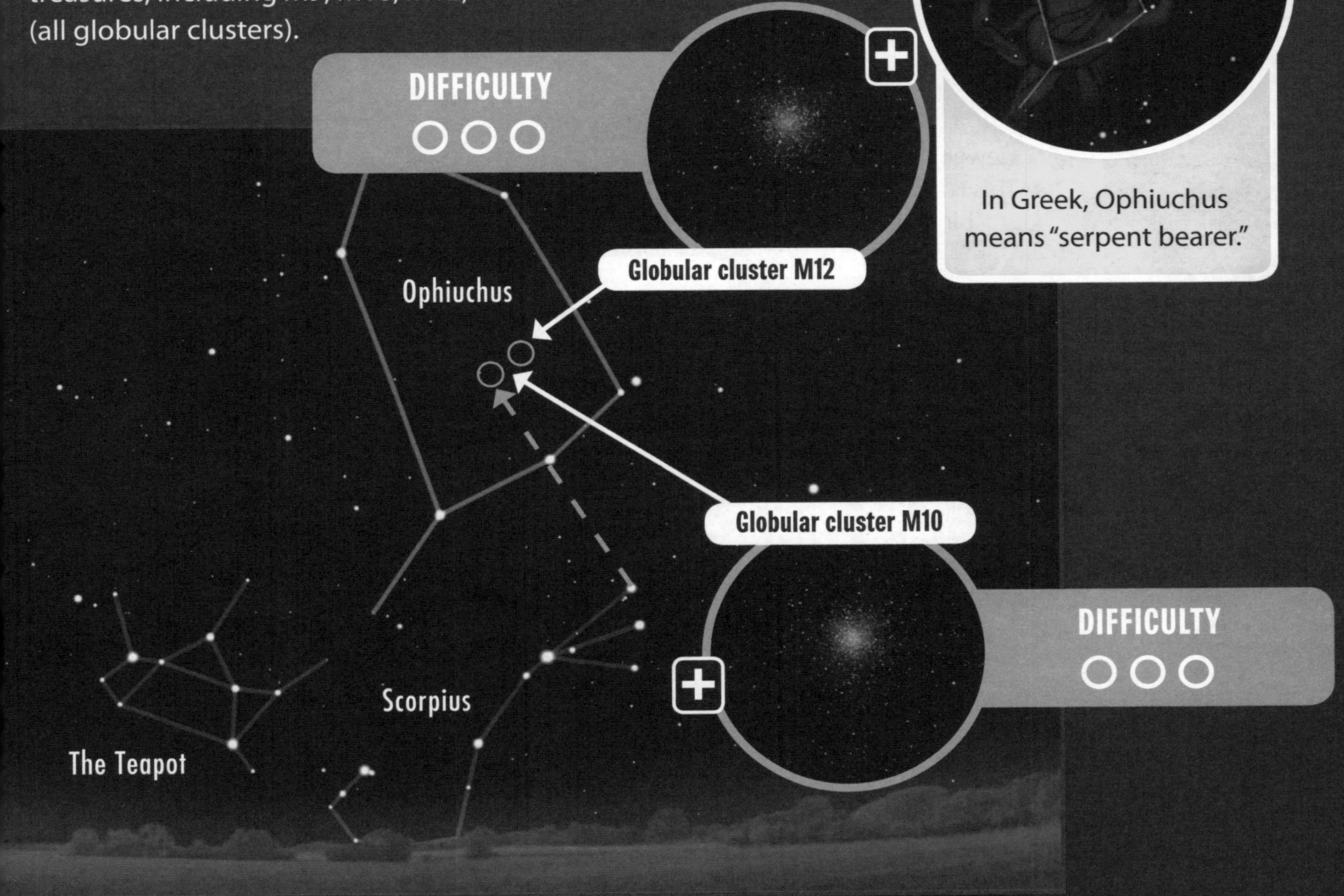

29 The Wild Duck Cluster (M11) in Aquila

Like Orion during the winter, Aquila (the Eagle), a summer and fall constellation, is recognized by an alignment of three bright stars. The brightest is Altair, which is bordered by Tarazed and Alshain. The Wild Duck Cluster is found near the Eagle's tail.

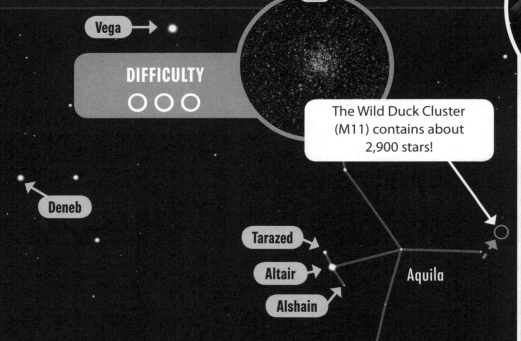

Vega →

DIFFICULTY
○ ○ ○

The Wild Duck Cluster (M11) contains about 2,900 stars!

Deneb

Tarazed

Altair

Alshain

Aquila

30 The Eagle Nebula (M16)

The Eagle Nebula is the source of one of the Hubble Space Telescope's most famous images: the Pillars of Creation (right). The nebula is in the constellation Scutum, but this is quite a dim constellation, so you may need to use the stars in the Teapot and Aquila as a guide.

DIFFICULTY
○ ○ ○

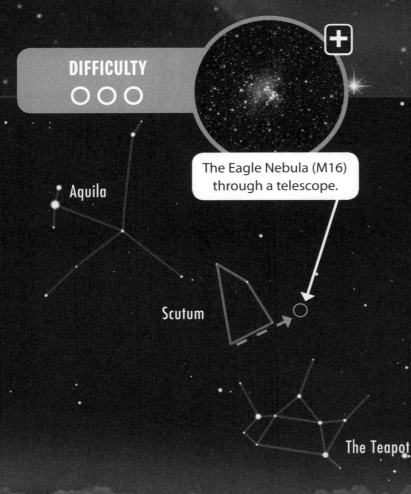

The Eagle Nebula (M16) through a telescope.

Aquila

Scutum

The Teapot

Stellar Facts

Scutum means "shield" in Latin.

31 The Ring Nebula (M57) in Lyra

Lyra, named for a musical instrument called the lyre (a small harp), is easily recognizable in the summer and fall skies by Vega, one of the brightest stars in the sky, and the diamond star pattern that makes up the rest of the constellation.

Lyra, the Lyre

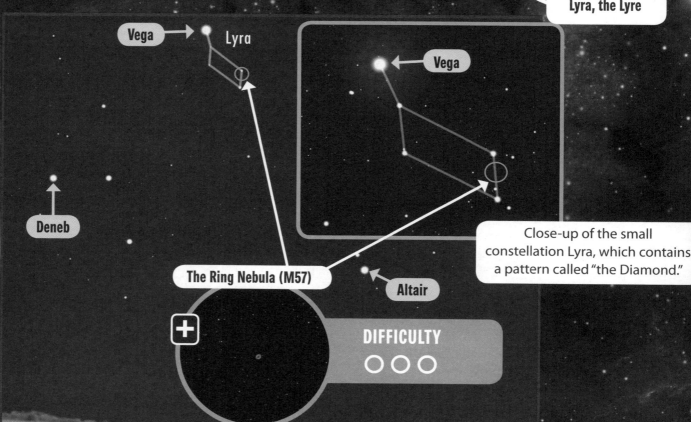

Vega →

Lyra

Vega

Deneb

The Ring Nebula (M57)

Altair

Close-up of the small constellation Lyra, which contains a pattern called "the Diamond."

DIFFICULTY
○ ○ ○

32 The Northern Cross, Albireo & Cluster M56

The Northern Cross is a star pattern located within the constellation Cygnus (the Swan). The brightest star is Deneb, which is also part of the Summer Triangle. Globular cluster M56 and Albireo are interesting objects to look at and can be found at the base of the Northern Cross.

Cygnus, the Swan

Vega

The Northern Cross

DIFFICULTY
○ ○ ○

This small globular cluster, named M56, is visible though a telescope in very dark skies.

Deneb

Altair

Albireo is a double star. The brighter star Albireo A appears yellow or amber in colour, whereas Albireo B appears blue.

DIFFICULTY
○ ○ ○

33 The Summer Triangle & Cluster M71

When you are able to identify Lyra, Cygnus and Aquila, it's time to put their brightest stars together in a star pattern known as the Summer Triangle. This combination helps you navigate the night sky and locate several interesting telescope targets including Sagitta and globular cluster M71.

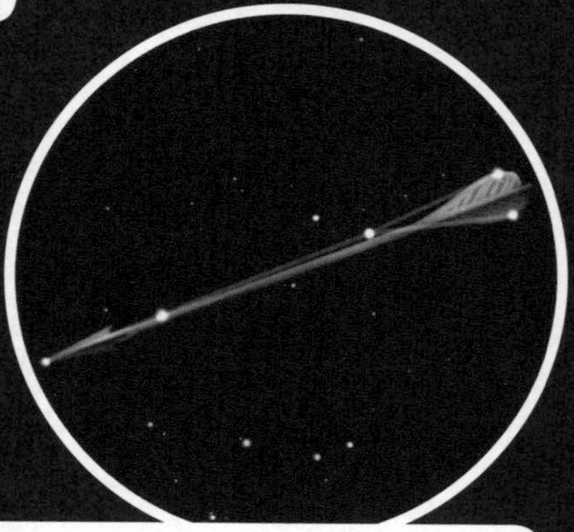

Close-up of the small constellation Sagitta, "the arrow."

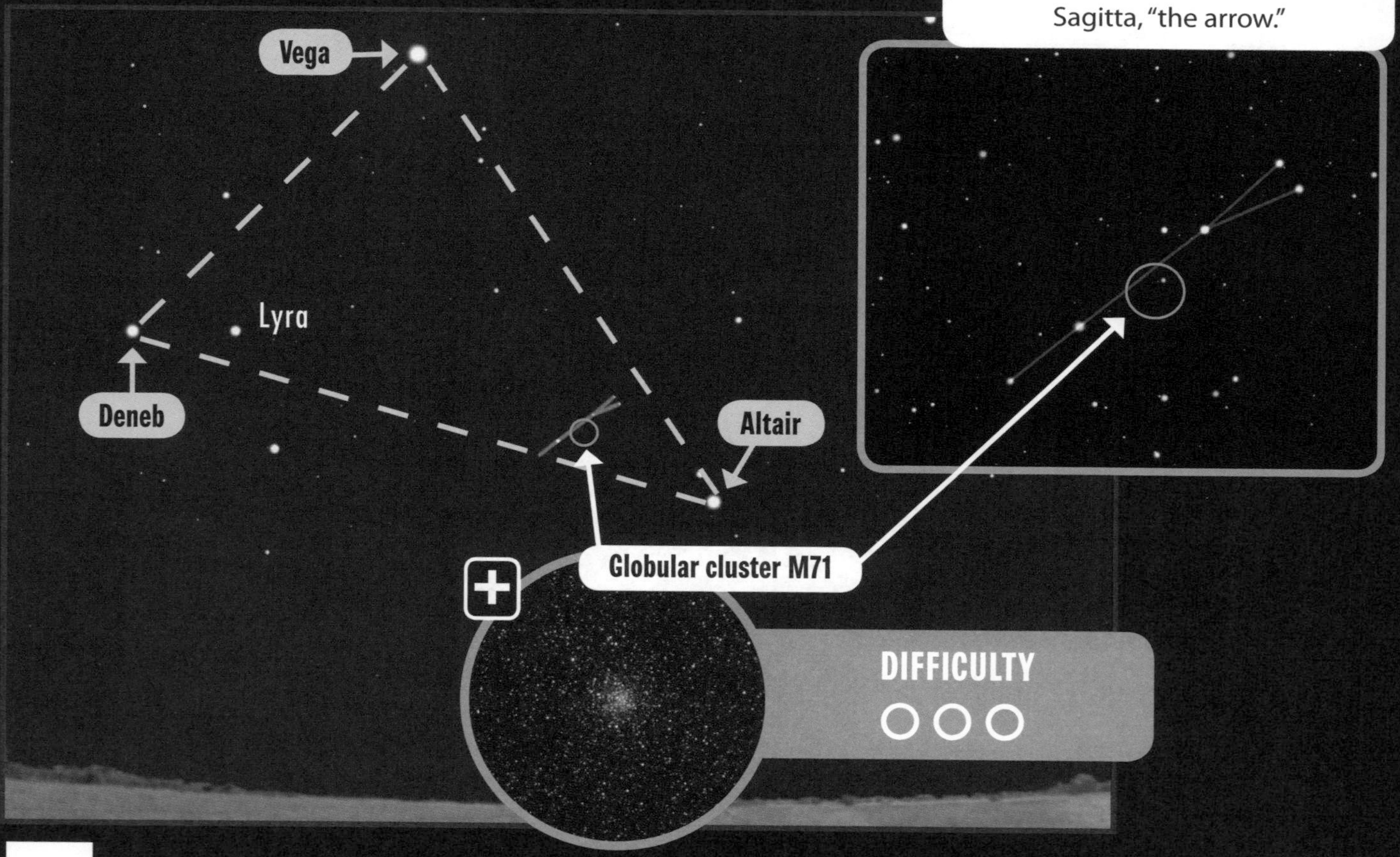

Vega

Lyra

Deneb

Altair

Globular cluster M71

DIFFICULTY

34 The Dumbbell Nebula (M27)

The Dumbbell Nebula is a cloud of glowing gas released by a star. It was the first nebula ever to be discovered. This nebula is so large, it would fill the space between our Sun and the nearest star nearly 4.5 light-years away!

Stellar Facts

Nebula means "cloud" in Latin.

DIFFICULTY

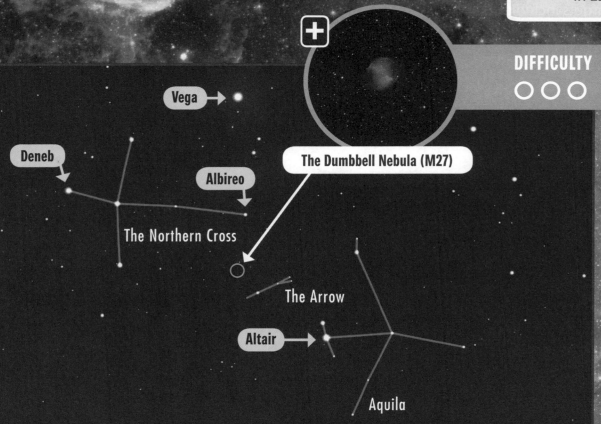

Vega

Deneb

Albireo

The Dumbbell Nebula (M27)

The Northern Cross

The Arrow

Altair

Aquila

35 The Coathanger

The Coathanger is a star cluster that rests right on the Summer Triangle. Look for six bright stars lined up in a row, with an additional four stars making up the hook in the hanger.

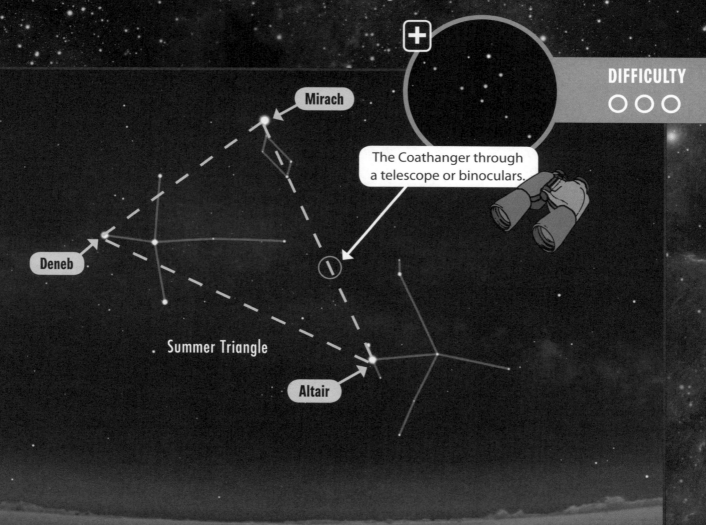

The Coathanger through a telescope or binoculars.

Mirach

Deneb

Altair

Summer Triangle

36 Cluster M13 in Hercules

The Keystone is an asterism within the summer constellation Hercules. This constellation is dimmer than most of the others, so it helps to find it in reference to other objects, like the bright star, Vega.

Stellar Facts

In Greek mythology, Hercules is the immortal son of Zeus.

The Keystone (asterism)

Vega

The Great Globular Cluster in Hercules (M13) is one of the brightest globular clusters. You will find it between two corners of the Keystone.

Can you still see the Summer Triangle?

DIFFICULTY
○ ○ ○

37 The Summer Beehive

The Summer Beehive is an open cluster located 1,400 light-years from Earth. Its collection of bright stars makes a wonderful target for binoculars or a small telescope. When observing this cluster, many people see the word "HI" written in the stars. What do you see?

DIFFICULTY

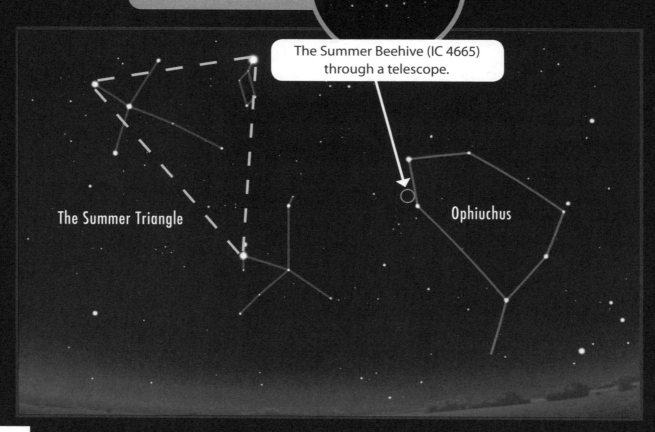

The Summer Beehive (IC 4665) through a telescope.

The Summer Triangle

Ophiuchus

CHAPTER 5
Autumn Objects

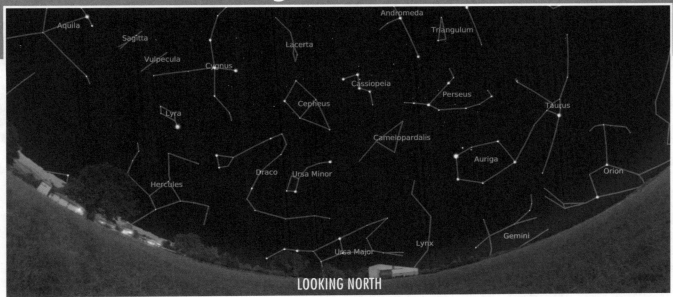

LOOKING NORTH

LOOKING SOUTH

38 The Triangulum Galaxy in Pegasus

Pegasus is located on the opposite side of Cassiopeia from the North Star. The Box is a pattern of stars that is part of the constellation. While the Box itself is devoid of popular stargazing targets, it can be used as a guide to find nearby targets, such as the Triangulum and Andromeda galaxies.

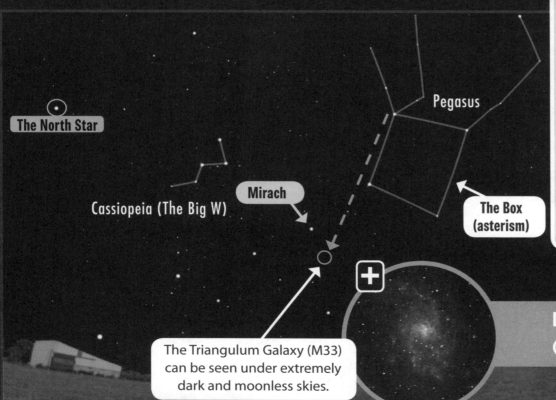

The North Star

Cassiopeia (The Big W)

Mirach

Pegasus

The Box (asterism)

The Triangulum Galaxy (M33) can be seen under extremely dark and moonless skies.

REMEMBER! Objects in the night sky appear to rotate around the North Star. Depending on the time of night, and time of year, Pegasus may appear to the right or left of the North Star, or above it.

DIFFICULTY
○ ○ ○

39 The Andromeda Galaxy

At only two million light-years away, the Andromeda Galaxy is the nearest galaxy to Earth (with the noted exception of dwarf galaxies, those galaxies with less than a few tens-of-billions of stars). In dark skies, this galaxy is visible even without a telescope. To find it, use the Big W to identify reference stars in the constellation Andromeda.

Stellar Facts

As summer turns into autumn, the days get shorter and the stars come out earlier. For this reason, many of the summer constellations can be seen well into the fall.

The North Star

The Big W

The Andromeda Galaxy (M31) through a small telescope.

Andromeda

This galaxy is a great target for binoculars, too!

DIFFICULTY
○ ○ ○

40 The Double Cluster

These two clusters, designated NGC 869 and NGC 884, are visible to the naked eye in extremely dark skies. However, through a telescope they are a magnificent sight in almost all conditions.

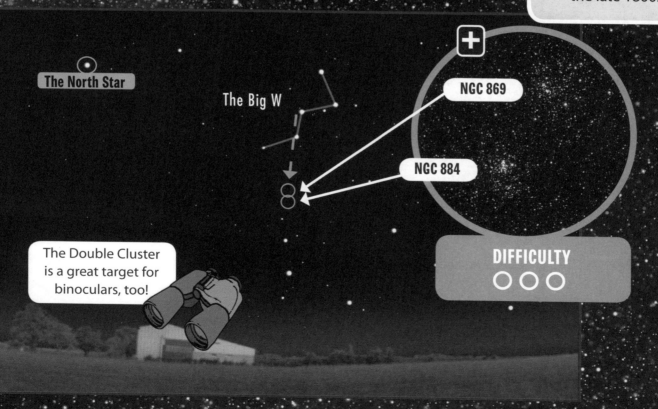

The North Star

The Big W

NGC 869

NGC 884

The Double Cluster is a great target for binoculars, too!

DIFFICULTY
○ ○ ○

60

41 The Iris Nebula in Cepheus

The Iris Nebula is a cloud of dust reflecting light from a nearby star. With dusty blue petals of light stretching over six light-years, this nebula is named after the flower. It was discovered by Sir William Herschel who also discovered the planet Uranus. Viewing this nebula requires extremely dark skies.

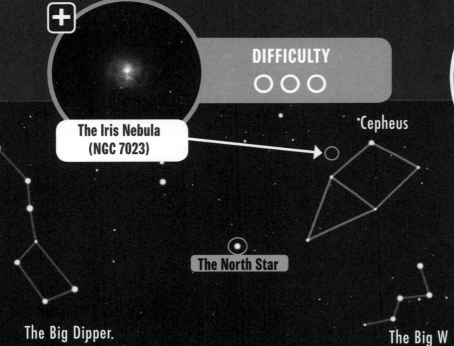

DIFFICULTY

○ ○ ○

The Iris Nebula (NGC 7023)

Cepheus

The North Star

The Big Dipper.

The Big W

CHAPTER 6
Other Objects

42 Mercury

Due to Mercury's extreme proximity to the Sun, it can be challenging to get a good look at it. It may only appear in the evening sky a few days per year. Because this planet is closer to the Sun than Earth, careful and frequent observation will reveal that Mercury has phases like the Moon!

Stellar Facts

In astronomy, Mercury and Venus are called the inferior planets because they are closer to the Sun than Earth.

This image of Mercury was taken during a flyby by the unmanned NASA spacecraft named *MESSENGER*.

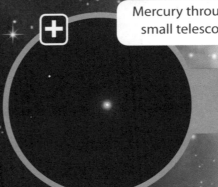

Mercury through a small telescope.

DIFFICULTY

○ ○ ○

43 Venus

Because it's so close to the Sun, Venus only appears shortly after sunset and shortly before sunrise. Just like Mercury, this planet is closer to the Sun than Earth and has phases like the Moon. Because Venus appears white through a telescope, some people momentarily think they're looking at the Moon.

Stellar Facts

Venus is named after the Roman goddess of love and beauty.

DIFFICULTY
○ ○ ○

Venus as it appears through a small telescope (notice how it almost looks like the Moon?).

This image of Venus was taken by the unmanned NASA spacecraft named *Mariner 10*.

While the average surface temperature on Mars is around -55 degrees Celsius, temperatures around the equator can rise to around 20 degrees. With a day only 37 minutes longer than on Earth, Mars is considered a prime location for human exploration. But humans will always need a spacesuit on Mars — it has just one per cent the atmospheric pressure of Earth.

NASA has been operating robotic rovers on the surface of Mars for the last 20 years.

NASA's *Curiosity* Rover

Mars is the fourth planet from the Sun and roughly a six-month journey via spaceship. However, to date only robotic spacecraft have visited.

Under ideal conditions, you may be able to see the polar ice caps and varying hues of red and brown. However, most of the time, through a telescope, Mars will look like a bright red star.

DIFFICULTY
○ ○ ○

Mars through a telescope.

45 Jupiter

Jupiter is the largest planet in the Solar System. It is bigger than all of the other planets put together. Jupiter's four brightest moons, discovered by Galileo Galilei in 1610, are visible through even the smallest telescopes and binoculars, too. Through your telescope, you should also see at least two cloud belts. If you're lucky and have a slightly larger telescope, you may be able to see the "Great Red Spot."

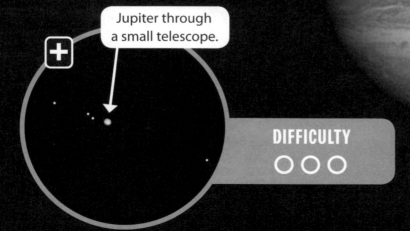

Jupiter through a small telescope.

DIFFICULTY

○ ○ ○

The Great Red Spot is a storm that has been raging on Jupiter for hundreds of years.

46 The Galilean Moons

Jupiter's four largest moons (called the Galilean Moons) change position every night, so you'll need to use astronomy software to help you determine which moon is which.

Jupiter's moons

DIFFICULTY

Ganymede is the largest moon in the solar system, having more than twice the mass of Earth's moon.

Europa is the smallest of the four Galilean moons. Latest estimates project that beneath an icy surface, there is an ocean 100 kilometres deep.

Callisto has the lowest radiation levels of Jupiter's large moons and would make a promising location for human settlement.

Io orbits most closely to Jupiter. It sports more than 400 active volcanoes! Due to the amount of volcanic activity, Io's surface features frequently change. Io has almost no meteor craters because lava fills them soon after they are formed.

Saturn is probably the single most fantastic thing that can be seen through a small telescope. Its majestic gold and brown hue is enough to take one's breath away.

The most spectacular thing about Saturn is its rings. Visible in even the simplest telescopes, the rings are made of mostly of ice. With a slightly larger telescope, you may be able to see a gap in the rings. This gap is called the Cassini Division.

At its closest, Saturn is over one billion kilometres from Earth. NASA's latest Saturn probe took six years and nine months to reach this planet.

Saturn through a telescope.

Saturn's moons

DIFFICULTY
○ ○ ○

Cassini Division

On most nights, you should be able to see Saturn's largest moon, Titan. But on really clear nights or with larger telescopes, you should be able to see several other moons, such as Rhea, Dione and Tethys.

This image of Saturn was taken during a flyby by the unmanned NASA spacecraft named *Cassini*.

48 Uranus

Despite the lack of visible details, the ice giant Uranus has several interesting features. This planet has 13 narrow, yet distinct rings. (These rings are only visible in professional telescopes like the Hubble Space Telescope.)

Uranus also has 27 known moons. If you have a fairly large telescope, you may be able to see the brightest five: Titania, Oberon, Ariel, Umbriel and Miranda.

Despite it looking very small through the average •telescope, you should be able to make out the distinct blue–green disk.

DIFFICULTY

○ ○ ○

49 Neptune

At its closest, Neptune is a whopping 4.3 billion kilometres distant. That's four hours at the speed of light! Its orbit is so wide that it takes 165 years for this planet to orbit the Sun. Neptune has an average temperature of -214 degrees Celsius and Uranus averages -216 Celsius. This is why these planets are nicknamed the "ice giants."

Neptune has 13 known moons. The largest is named Triton.

Through a telescope, Neptune is clearly blue. You should also be able see Triton, Neptune's largest moon

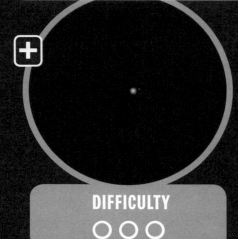

DIFFICULTY

○ ○ ○

You might be wondering why Pluto doesn't have its own page in this book. That's because Pluto is so small and so dim that it is extremely challenging for amateur astronomers to observe.

50 Comets

Nicknamed "dirty snowballs," comets are made mainly of ice and dust. When comets are near the Sun, they release gas and dust in the form of a long (and sometimes colourful) tail. When a bright comet is visible from Earth, it usually makes the news. They can be difficult to see clearly with the naked eye, so it helps to use binoculars or a telescope.

Comets are great targets for binoculars, too!

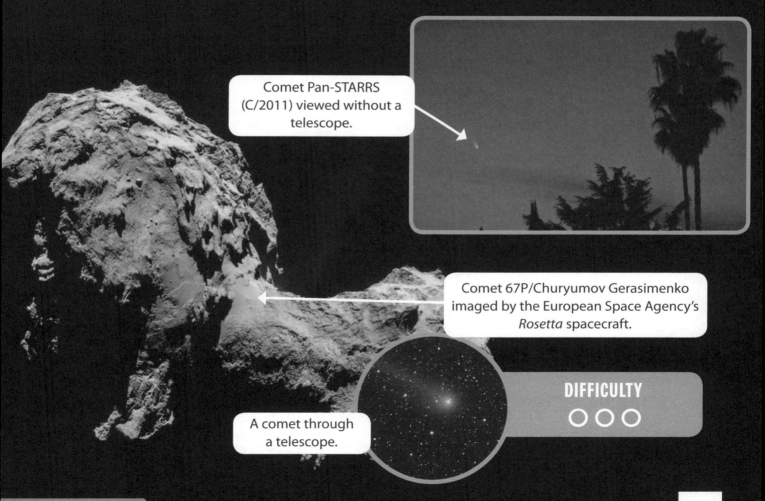

Comet Pan-STARRS (C/2011) viewed without a telescope.

Comet 67P/Churyumov Gerasimenko imaged by the European Space Agency's *Rosetta* spacecraft.

A comet through a telescope.

DIFFICULTY
○ ○ ○

Glossary

ASTEROIDS: Space rocks that orbit the sun. These are found mainly between Mars and Jupiter in the "asteroid belt."

ASTERISM: A pattern of stars within a constellation. Asterisms have handy names like the Big Dipper in the constellation Ursa Major, the Keystone in Hercules, or Orion's Belt in Orion.

ASTRONOMY CLUB: A place where people gather to appreciate the wonders of the night sky. Astronomy clubs often host stargazing parties and lectures from prominent scientists. To find the club nearest you, visit: www.skyandtelescope.com/astronomy-clubs-organizations

CIRCUMPOLAR: Objects in the night sky that are always above the horizon and appear to circle the North Star (in the Northern Hemisphere).

CONSTELLATION: A group of stars forming a recognizable pattern. To help classify stars by location, ancient astronomers divided the sky into 88 distinct regions called constellations. They were named after mythical creatures and heroes.

DARK SKIES: Moonless, cloudless nights far from city lights. To find the dark skies nearest you, visit: darksitefinder.com/dark-sites

DEEP-SKY OBJECT: A stargazing target that resides outside our solar system. This includes galaxies, nebulae, globular clusters and open clusters. Deep-sky objects are tens of thousands, or even millions, of light-years away!

ECLIPTIC: The apparent path the Sun takes through the sky throughout the year. The Moon and planets are always found near the ecliptic.

GALAXIES: Clusters of millions, billions or even trillions of stars. Our home galaxy, the Milky Way, is visible as a cloud of stars that crosses the entire night sky.

GLOBULAR CLUSTER: A tight group of thousands of stars orbiting our galaxy in a region called the Halo. The closest globular clusters to Earth (M4 and NGC 6397) are 7,200 light-years away.

LIGHT-YEAR: The distance light travels in one year. It is used to measure distances in space. One light-year is about 9,500,000,000,000 kilometres.

MESSIER LIST: The most interesting (and easiest to find) deep-sky objects compiled by a French astronomer named Charles Messier.

NEW GENERAL CATALOGUE (NGC): A list of almost 8,000 nebulae and star clusters compiled in the late 1800s.

NEBULA: A giant cloud of gas and dust. Some nebulae are formed after a supernova, when a star explodes. Others are formed when a smaller star blows off its outer layers near the end of its life. Nebulae are also the place where new stars are formed.

OPEN CLUSTER: A group of stars that formed around the same time.

ORBIT: The curved path an object such as a planet, moon or spacecraft takes as it travels through space. Most orbits we observe are elliptical, such as the path the Earth takes around the Sun. Planets orbit stars and moons orbit planets.

PLANETARY NEBULA: A nebula formed during the aging process of older stars.

PLANETS: Massive objects (often called worlds) that orbit a star (as opposed to moons which orbit planets). To be considered a planet, a world must be massive enough to be roughly spherical and have cleared its orbit of debris.

STAR: A massive ball of hot gas that produces energy by nuclear reactions. Our Sun is a star, as are most of the specks of light appearing in the night sky.

SOLAR SYSTEM: The region of space that includes our Sun along with the eight planets, dwarf planets, asteroids, comets and everything else that revolves around it.